Expressive Arts Your Way

A WORKBOOK FOR CHILDREN AND ADULTS

JOYA GRANBERY HOYT

©2022 Joya Granbery Hoyt

All rights reserved. No portion of this book may be reproduced, mechanically, electronically, or by any other means, including photocopying, without written permission of the author.

Edited by Terri Zumstein
Cover Dersign: Interior Design and Layout: Terri Zumstein

terri@qualityinstantprinting.com

DEDICATION

This Book is dedicated to my sister Pamela Granbery - my partner in SISTERS Dance Company, OUTSIDE THE BOX and all the other creative endeavors represented here. Without her imagination, encouragement and hands on hard work, none of these projects would have happened.

Special thanks go to Martha Deichler, BSUSD, Terri Zumstein at Quality Instant Printing, Kiwanis, Rotary Borrego Springs, St. Barnabas Church, Borrego Springs Chamber of Commerce, Borrego Sun Newspaper, Borrego Art Guild, Borrego Art Institute and Christopher and Katherine Carbone at the Salve Regina University Expressive Arts program.

MISSION STATEMENT

The object of this book is to suggest and provide a playing field for a series of exercises that can be fun as well as healing backgrounds. In addition to variety, the object is to limit the number of supplies necessary to accommodate limited budgets if any and supplies that are readily available in any environment. With a few exceptions, these classes are all designed to take place at the participant's location.

INTRODUCTION

This workbook is designed to be used in a number of different ways.

It can be referenced as an idea book for individual workshops. It can be assembled in an order that would provide a series of activities in a specific pattern like from easiest to most difficult.

It can be set up so that whole genres are removed from more site specific venues.

Thus it is more adaptable than a fixed agenda would be. I would insert empty pages for today's session with the assignment at the top.

For the purposes of this course work in group applications, I have included most of the suggested activities in an order that moves from Middle School to the general public to more esoteric adult material. Warm-ups and closure exercises at the beginning and end are suitable for all groups. Lesson plans for school children can also be adapted to adults and vice versa.

LESSON PLANS: *Middle School*

Sewing

Cooking

Competition Dance

Dance as Exercise

Ikebana

Film Shorts

Motivational Cards

Alcohol Ink Painting

Rain Painting

Mandalas

Percussion

Public Exhibits

Field Trips

Founder Joya Granbery Hoyt

Lesson Plan - Middle School Sewing Classes

Sewing Projects
Learn how to operate a sewing machine.

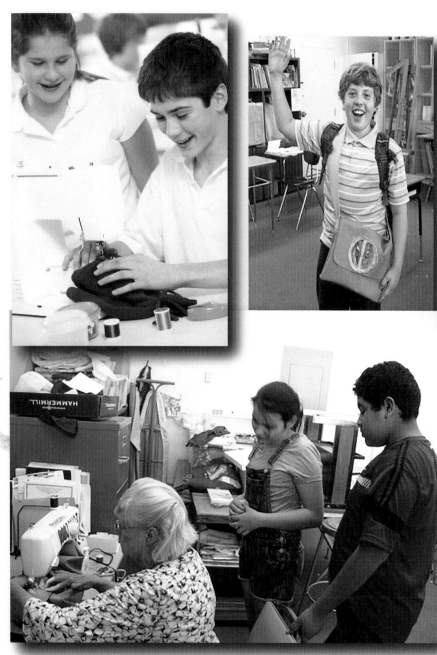

Sewing is successful because projects and fabrics have appeal to modern kids. A variety and quality of projects keeps them interested and coming back for more. Kids learn a valuable skill that they can use later in life.

Book bags, pajama pants and pillows are simple items to begin with initially. Patterns are easy to find. You can make multiple patterns by copying them onto newspaper or shelf paper so everyone has one.

Learning The Art of Sewing: Beginner Basics

Students can begin their creative journey learning the art of sewing with understanding machine basics including machine anatomy, understanding various types of fabric, and stitches and terminology.

Learning the basic stitches, and when to use them as well as learning how to sew in a straight line are the initial tasks.

Sewing is an art that serves as an essential tool and creative outlet for everyone who learns it.

Sewing - group quilt project

The cut out collage designs can be transferred onto fabric and sewn together to make a quilt. This quilt was made in a sixth grade art class and hangs in the community room at the high school. It was auctioned off at a benefit to raise the money for the next year's program and then donated to the school.

COLLAGE QUILT

Quilt displayed at the BAI before the auction where it sold for $1200.

If you want to design your own clothes or costumes but don't feel comfortable drawing fashion illustrations, cut out a figure from a magazine and trace around it on a blank piece of paper. Now you have a mannequin to design your fashions on. You don't have to be able to draw human figures yourself to start designing wardrobe.

* *The costume for the lizard bird above is an example of a costume sketch on a cutout model from a fashion magazine as mentioned above. You can also have someone draw a line on the floor around your own body in an interesting shape on a roll of paper and design wardrobe on that.*

* *Costume for lizard-bird for dance film FROZEN STREAM - can be seen on you tube under Joya Granbery Hoyt – all these films are appropriate for families and schools*

Frozen Stream

Dance as Exercise

Lesson Plan - Middle School Dance as Exercise

Learning, thought, creativity, and intelligence don't just come from the brain alone, but from the entire body. Movement combinations increase memory, order, and sequencing skills. Creating dances also increases self-esteem which is so very important to learning. We already witness the need for children to move throughout the day. The positive effects that music and dance have on students' development is extraordinary.

Swirling, billowing, flowing scarves are perfect for extending range of motion and are one of the best ways to develop creative large motor skills in even your most shy students. All children love to pretend they are birds, super-heroes, fish, statues, planets or anything. Bright colors improve visual tracking, while the free-flowing nylon allows all types of non-verbal play and creativity.

Dance warm-ups can get the energy flowing or they can calm people down.

Ballroom dance music and steps as well as abstract modern dance movements bring visual and physical excitement. If students are experienced or trained dancers, it's fun to do creative movement pieces with them. You can add lengths of fabric that can be used as capes or wings or to wrap yourselves in bright colors to add to the excitement. Adding music can stimulate or calm down the group.

Look for the expressive dance film WIND SEA SKY on you tube under Joya Granbery Hoyt - all are appropriate for families and schools.

Lesson Plan - Middle School Competition Dance

Dance is a terrific socialization exercise as well as a confidence builder

The ability to positively impact the community is one of the biggest reasons to teach dance in schools. Children learn to express their emotions through movement, and the focus that dance demands can help them find greater stability in their lives and form stronger self-identities. On a larger level, students create strong bonds with one another through dance, and parents and faculty are also connected through the dance activities of their children and students.

The group of sixth graders above were the first class to actually dance together ballroom style at their junior prom. Sixth grade students can learn the cha cha and samba using weights for dips since they are not allowed to touch each other in Middle School. Once they achieve the team level they can compete at area colleges against college teams.

Lesson Plan - Middle School Cooking Classes

Benefits of Cooking Lessons

Engaging lessons will have students developing skills and thinking about learning in a new fashion that is fun and exciting for everyone involved.

1) Children may *try new and healthy foods*. Recent research published in the Journal of the Academy of Nutrition and Dietetics indicates that children engaged in tactile experiences, such as handling food have a greater acceptance of eating a variety of foods.

2) A *kitchen is a learning lab for children* that can involve all of their senses. While kneading, tossing, pouring, smelling, cutting, and feeling foods they have fun and learn without being aware of it.

3) Children who cook at home indicate a *"sense of accomplishment," self-confidence, and feeling of contributing to their families*.

Children tend to *skip less healthy prepared or processed snack foods as they prepare their own food more*. Recent research indicates that nutrition knowledge may be incomplete without the experiential learning or hands-on activities associated with food preparation that involves handling food and cooking equipment.

Children *learn lifetime skills through practicing basic math skills such as counting, weighing, measuring, tracking time*.

They also gain social skills by working together and communicating in the kitchen. Teaching cooking to youth is an opportunity to teach nutrition education such as planning meals and make smarter food choices for themselves and their families.

Simple cooking classes can produce snacks to cater an art opening at school and foster student interaction with their community.

Below are simple recipes from Chef Pam's 6th grade Clean Hands Catering

HERBED CREAM CHEESE ON CUCUMBER ROUNDS

INGREDIENTS
3 lb. cream cheese
2 1/2 cups sour cream
1/4 c. blue cheese
1 cup fresh and dried herbs
Salt and pepper
3 or 4 cucumbers

INSTRUCTIONS
Mix all cheeses and sour cream in blender
Chop herbs and add to cheeses
Add salt and pepper
Score cucumber with fork or peeler
Slice into thin rounds
Spoon cheese mixture onto cucumber round
garnish with sprig of fresh herbs

ROSEMARY WALNUTS

INGREDIENTS
2 c. nuts (walnuts, almonds, pepitas etc)
2 1/2 Tab butter
2 tsp rosemary,
1 tsp salt,
1/2 tsp cayenne

INSTRUCTIONS
Melt butter and spices
Pour over nuts
Bake at 350 for 10 minutes

More KIDS RECIPES

BACON WRAPPED DATES

INGREDIENTS
60 dates
2 lbs. bacon
60 toothpicks

INSTRUCTIONS

Pit all 60 dates
Slice bacon into thirds
Wrap each date around one third bacon slice
Spear with toothpick
Bake at 425 degrees for 10 minutes
When crisp turn if necessary

SWEET POTATO DIP

INGREDIENTS
5 sweet potatoes
2-3 limes and their juice
2 chipotle peppers in adobo sauce

INSTRUCTIONS

Cook sweet potatoes at 350 in oven for 1 hour
Zest and juice limes
Seed and mash peppers
Pinch of salt
Mix potato, zest, juice and one pepper in blender
Taste for salt and heat of peppers
Garnish with extra zest
Serve with vegetables or pita
or homemade tortilla chips

HIBISCUS LEMONADE

INGREDIENTS
3 cups sugar
3 cups lemon juice
water
ice
lemon slices,
1 cups hibiscus tissane

INSTRUCTIONS

Mix all ingredients
Stir until sugar is dissolved
Add hibiscus tea if desired
Serve over ice
Garnish with lemon slices & mint

Lesson Plan - Middle School Ikebana Art of Arranging Flowers & Branches

The Japanese art of flower arranging, blossoms, branches, leaves, and stems find new life as materials for artmaking. In contrast to the western habits of casually placing flowers in a vase, ikebana aims to bring out the inner qualities of flowers and other live materials and express emotion.

You may even paint the leaves of an element. Plant limbs may be arranged to sprout into space in various directions, but in the end, the whole work should be balanced and contained. Every arrangement takes on a sculpture-like appearance.

Students can use items found outside with no need for flowers Items such as dried leaves, branches, stones, dead flowers, and feathers arranged in a vase from the dollar store can create a unique floral arrangement.

Lesson Plan - Middle School Mandalas

Mandalas are spiritual circle designs that you can make by yourself or with someone else or with a group. They can be simple drawings in the sand or elaborate collages.

In its simplest form, a mandala is a circular structure with a design that radiates out symmetrically from the center. You can find natural mandalas in flowers, tree rings, the sun, eyes, snowflakes, spider webs, sea shells, seeds, fruits and more.

MAKING A NATURE MANDALA

Pick a peaceful place out in nature to create your nature mandala.

Then, you will need to gather some organic materials around you to use.

To create your nature mandala, place a meaningful item in the center.

Continue making patterns until your items are used up and your nature mandala looks complete.

Many mandalas are made with geometric diagrams and colored using pencils or markers, but mandalas made in nature are simplistic and easy to create. An elaborate sun, a circle filled with stars within stars, colorful bits of paper, a simple spider web--all mandalas each have an attractive pattern that captures one's attention and stimulate your imagination as a myriad of colors are displayed. Hours of peaceful pleasure follow.

Mandala in the snow with a circle of seaweed

Lesson Plan - Middle School   Film Shorts on Cell Phones

A short film is any motion picture that is short enough in running time not to be considered a feature film.

Students write, produce and film short subjects on their cell phones which are shown at local Film Festivals.

A short film is an inexpensive way to learn the production process, and gain experience. A short film can be looked at like a trial. If your short film is successful, it can lead to larger projects, higher budgets, and feature film production opportunities. Some people just love short films.

Where you can Watch Award Winning Short Films or live stream short films

1) Sofy.tv: A dedicated short film streaming platform.

2) YouTube: The General Purpose Video Streaming Site.

3) Filmsshort.com: Dedicated Platform Which Relies On YouTube or Vimeo.

4) Netflix: The Film And TV Streaming Site.

ShortsTV is the world's first and only 24/7 HD TV channel dedicated to short movies.

Available around the world, the channel brings you professionally produced live action, animation and documentary short movies, including hours of award winning and star-studded shorts. ShortsTV also offers original programming on the global short film scene, keeping you up to date with the latest events and putting the spotlight on both emerging and established talent.

ShortsTV is available in the US on the following platforms:

DirecTV – Channel 573

AT&T U-Verse – Channel 1789

Google Fiber – Channel 603

Hotwire – Channel 560

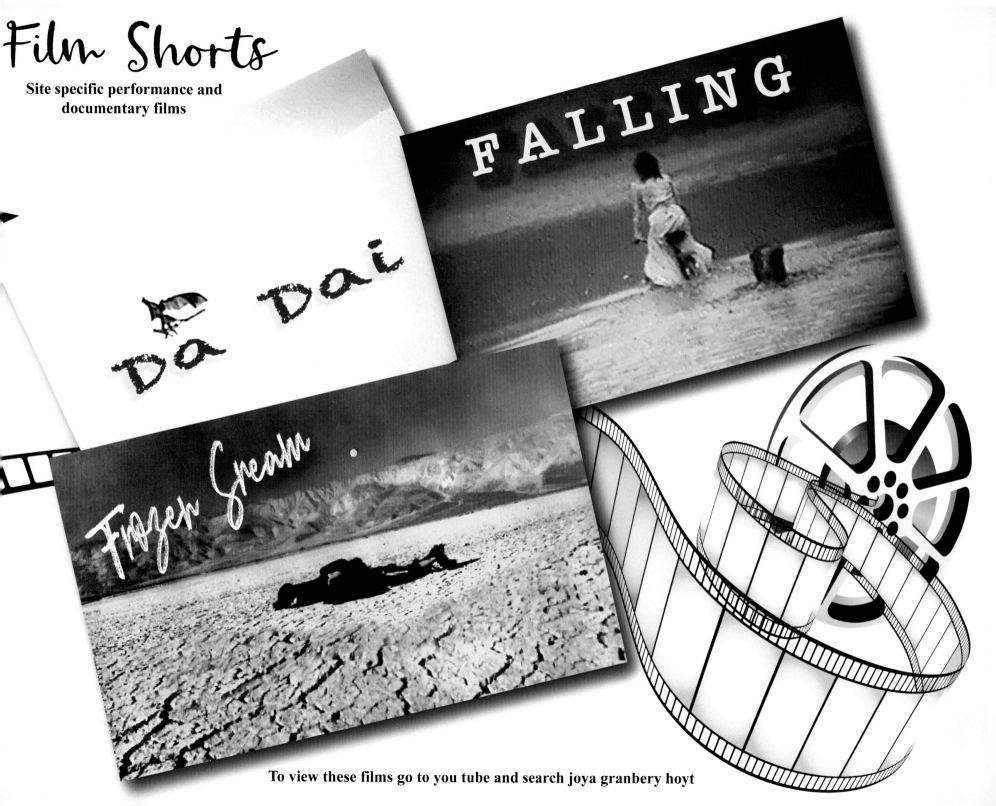

Lesson Plan - Middle School Motivational Cards

Motivational cards connect you into out-of-the-box thinking, magical imagining, and a deep connection with the creative spirit that is your deepest, truest self.

Motivational Cards are fun for all ages to make. You can select to motivate a project or just give you an inspiration for your day. It's fun to make a shelf or a display rack to show them and a special pouch or folder or box to keep them in.

In a warm-up exercise with motivational cards, participants select or make a playing card sized drawing of an image that is inspirational. The next step is to write a fill in the blank sentence of an image on the back. For example, "If today were my birthday I would like a (Blank)"

You can always start by picking a card from tarot or angel cards to use it as a guide to make your own card. Think about why you may have chosen that picture. Color the picture in with crayons or colored pencils. Now flip the card over and fill in the blank for the sentence chosen for today– for example, "If I were a bird today I would be a (blank)." See if you can imitate the kind of movement or sound that a bird would make.

These cards are very useful when you are having a hard time making a decision or lacking a sense of direction .

Folder to store cards in

Lesson Plan - Middle School Rain Painting

Turn A Rainy Day Into Art!

For a fun rain painting technique, take a large piece of any kind of paper.

Use water based markers, water color pastel paints, food coloring or leftover water based old house paint.

Use an eye dropper or a spoon or an old paint brush to drip paint all over the paper.

You might like to place your paints very close together, or space them out – or try both and see what different effects you can create. Be generous with your paints, and keep them liquid, adding plenty of water to blend the colors.

You'll get a more striking result if you color the whole page and overlap the colors

OUTDOOR RAIN PAINTING STUDIO

Ideally we can leave the painting out in the rain so the color run - otherwise we can spray the painting with a watering c or a hose. After we let the painting dry, we write our though about the results. Do we see any images emerge ? How does the painting make us feel ?

Rain Paintings

You can also use very inexpensive paper like mailing and moving company wrapping paper for large pieces

When the rain paintings are dry you may also want to draw on them with other colors and designs Rain paintings can make fun wrapping paper .

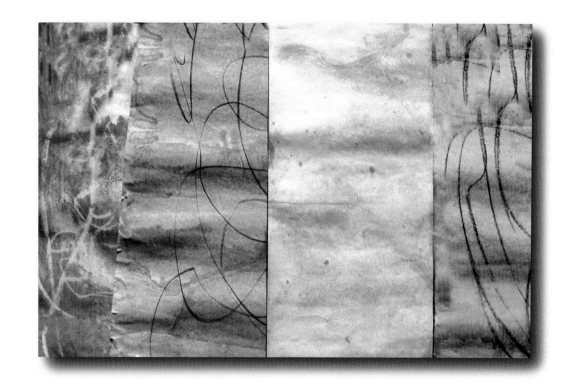

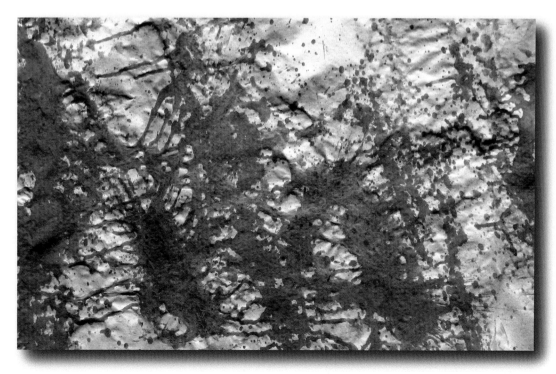

Sand Painting

Desert Field Trip

Lesson Plan - Middle School Alcohol Ink Painting

Splash, Tip and Apply Canned Air on Tiles

The vibrant, pigmented and user friendly alcohol inks lend themselves to explode on any non-porous substrate such as ceramic tile of all sizes and shapes, Yupo paper, metal, glass and wood to name a few. They dry very quickly and have a natural variation in color and texture. You simply splash, tip, blow through a straw or use canned air to create stunning art masterpieces. **There is no requisite for artistic talent.** In a thirty minute time span you can go from beginner to an accomplished artiste!

This medium affords its use for children as young as 4 years olds to autistic children and adults with disabilities as well. Because of no requirement for art skills and quick learning capabilities, students have a finished art piece that can make them feel they have accomplished a skill they can be proud of in an hour or less.

Tile painting class

4 year old child painted tile for jewelry box

Group of 14 students painted tiles in mirror above

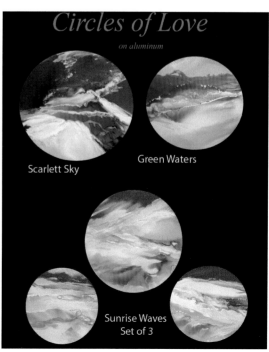

Aluminum circles

Lesson Plan - Middle School Percussion

Percussion musical instruments are those you play by hitting them with your hand or an object such as a stick: Drums, tambourines, and cymbals are all percussion instruments.

An easy way to make a percussion instrument is to take a tube and fill it with pebbles and tape over the ends - when you shake it you get a sound like an Indian rain stick.

Many exciting group music events can occur with groups and individuals who do not actually know how to play an instrument . This group of childhood music boxes on the right was used as an orchestra by a group of students who composed a score which indicated which instruments were to be played when - at some points they all played at once and then there were solos and duets - the piece ended with an old broken box playing a sad solo that was very haunting and dramatic.

When it comes to percussion instruments, most people tend to think of the drum. However, there are different types of percussion instruments from different parts of the world. At one point I had an opportunity to play in an orchestra conducted by John Cage and my instruments were a wad of paper and a telephone receiver . He complimented me on the texture of my paper playing. Anything that makes a sound is an instrument in this kind of orchestra. The trick is to decide who plays which instrument when. Anything that makes a sound can be an instrument. That includes your own voice.

Percussion instruments have been instrumental in adding color and rhythm to music over the years. Children are introduced to percussion instruments as infants and their popularity continues to make percussion instruments fan favorites for all their childhood and beyond.

Examples of percussion instruments include rain sticks we made above, as well as shakers, tambourines, maracas, xylophones, blocks, and bells.

Lesson Plan - Middle School

Public Exhibits of Student Works & Field Trips to Museums / Art Galleries

Students projects can be shown and sold at local art galleries, boutiques, pop-up stores in local malls and street fairs

Field Trips to Museums – You can arrange a visit to a local gallery or show. Be sure to have everyone vote on his or her favorite piece – the galleries love to put up a sign next to the work saying it was the students' choice for best in show and it's a way to make sure they look at the work carefully and pay attention to the purpose of the trip.

Pam Granbery

Joan Anson – BAI Tour Guide
Borrego Springs Art Institute (BAI)
in Borrego Springs, CA

Field Trip to Airplane Museum

LESSON PLANS: *Adults*

Collages

Mandalas

Self image drawings

Archetypes - drawings

Diagramming phone calls and conversations

Storytelling

Diagram sleep patterns and draw dreams

Color fields - inspirations from nature

Consciousness - nature

Designing personal work and living space

Drawing to Music

Free writing - how you feel

Breathing

A chart of your life from the beginning until now

Workspace

One of the first things you want to think about is where and when you will be working on your expressive arts projects large or small . Ideally "when " will be a time when you have some quiet time to yourself . "Where" is just as important . It can be as simple as a picnic table in the backyard or a small corner of a sun porch where you can write and draw . A board or hollow core door on sawhorses in the garage or storage unit can give you a place where you can leave work out to visit when you have a few extra minutes or leave wet projects to dry undisturbed . The back yard is a great environment for fresh air and lots of space to be messy.

I have also taken a cup of coffee down to the harbor and worked for hours on the dashboard of my car . I even drew an elaborate floor plan for a house that I could work in and drew sketches of the furniture that would be in my studio and figured out what kind of pick up truck I would need to be an artist . A sign outside said RAIN PAINTING EXPRESSIVE ARTS BY APPOINTMENT. In Mexico I got up at 5 am and started journaling while the neighbors would load their surf boards into their cars and drop down to the beach in the dawn light while I wrote in the window .

I have watched my mother in law and women in Cuba paint on the kitchen stove while cooking . My mother worked on her architecture projects on the dining room table while we were in school . What would the ideal shape for your dream studio be ? How about a circle ? I like the concept that the straight line is the downfall of mankind and everything the American Indian does he does in a circle . Try painting in a circular tent or an igloo.

Studio

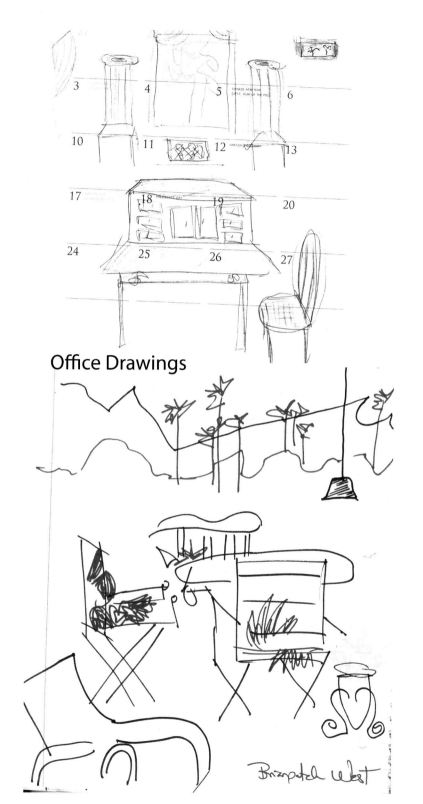

Office Drawings

Color Field

Inspiration from Nature

Color field painting expresses a yearning for transcendence and the infinite. These abstract ocean and desert scenes by Pamela Granbery were painted with acrylics on canvas with iridescent paints for sparkle and glow.

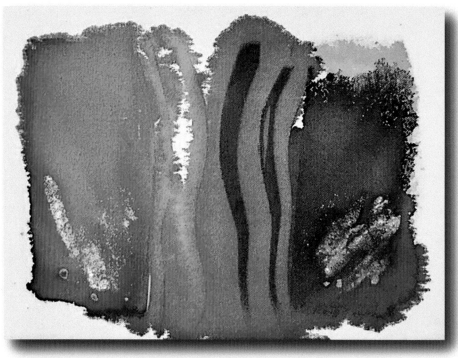

Inspiration from Nature by Pamela Granbery www.jessicahagengallery.com

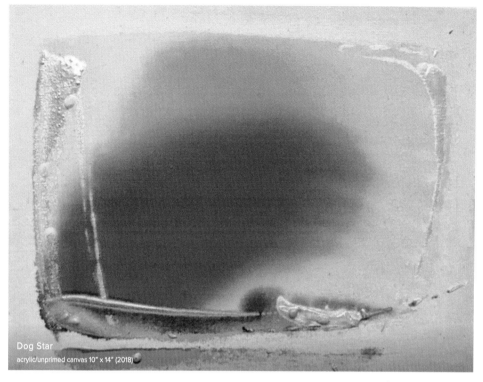

Dog Star
acrylic/unprimed canvas 10" x 14" (2018)

Mandalas and Collage *Design as a combination*

In section one we talked about collages when we did the Indian rug designs. We also talked about mandalas including out in nature - in this section we will explore the use of collage on mandala design as a combination. It's fun to get really involved with collage and circular collages are an exciting variation so working on a mandala base can be quite detailed and beautiful to look at. Don't hesitate to get a little carried away with this one - Making circles in the sand and decorating them or sand painting are other variations on this theme.

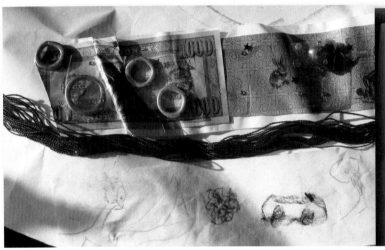

Life Drawing

Life drawing, also sometimes known as figure drawing or gesture drawing, is the drawing of the human form in various poses and levels of detail.

These drawings were done with magic markers which are colorful and easy to handle.

More advanced students can use charcoal or paints.

Storytelling in groups, individually and in performance

We all know the game where each person adds to a story going around the circle. Another fun way to tell a story is to give out the punch line to the storytellers so everyone's story has to end with the same phrase. In this picture I am telling a story that has to end with the phrase " Swipe Right ". I changed the dynamic by telling a sweet story about a little boy with a happy ending that surprised people who were expecting a dating joke.

Another exercise that tells a story is to pick out one of your favorite fairy tale illustrations and make up your own story about what is going on in the picture.

Story of Your Life . . . *from the beginning until now*

This is a 12 foot long mural which is a tapestry of the story of my life from the beginning until now.

Basically you take a roll of shelf paper with lots of feet left at the end for the future - maybe the whole roll! - start the timeline at the beginning with cartoons illustrating milestones in your lives until now. The post it notes are to identify areas that have some power in our memories and then write a paragraph about that later. The post it notes can be moved around on the scroll as different episodes bring up different reactions.

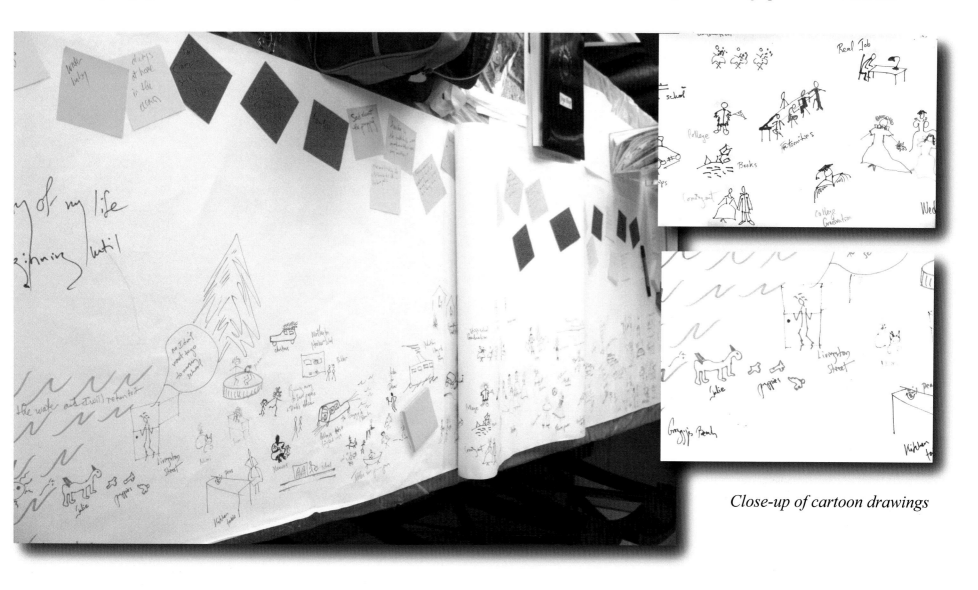

Close-up of cartoon drawings

Journaling

Creative journaling has been especially successful in helping people express their silent internal ingrained psyche because it uses both imagery and words, while at the same time, requiring practically no skill.

As you begin your own creative journey in expressive arts one of the first things you will want to do is keep a journal. I personally have three journals going all at once. I have a tiny notepad with me to jot down ideas and things I see or hear that I don't want to forget. My second journal is a lined writing notebook where I record larger passages of things I read or my ideas for projects I would like to do. I also tape or paste cut outs from magazines or newspapers into this journal as reference material for ideas I want to pursue later or just pictures I like. My third journal is a large book of plain paper which can be an artist's painting pad where I draw or paint pictures and designs for projects I might like to pursue more seriously like a portrait or a landscape or experiment with drawing to music or my own breathing.

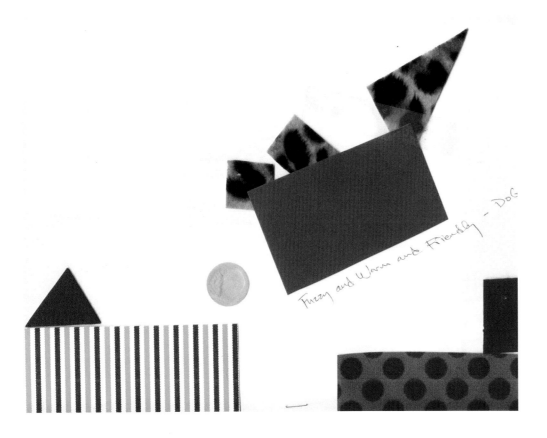

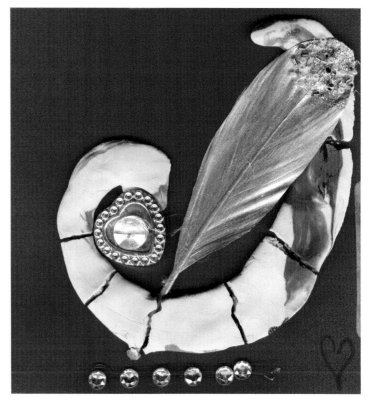

Dreams

You can do sketches of your dreams for instance. This can often help you figure out what to do about issues that concern you. In this stick drawing I am rushing to the airplane and my sister is tagging along behind with too many bags. Doing a drawing about it helps take the anxiety away.

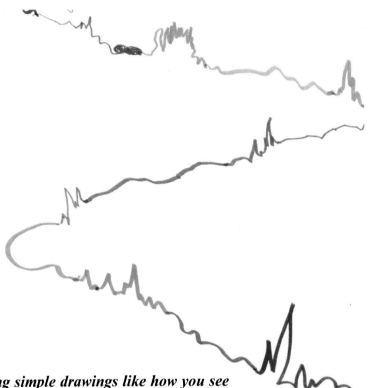

Practice doing simple drawings like how you see yourself physically and mentally in the moment. Have some fun with it - use color

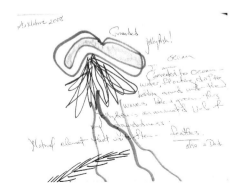

Emotions

You can do drawings of your emotions as you feel them come up and see what they feel like for you.

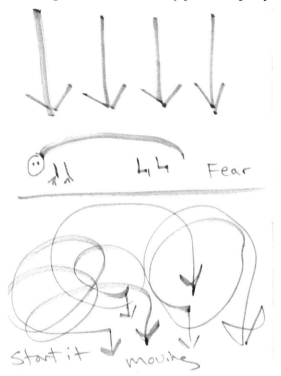

Diagraming Telephone Calls

One of my favorite exercises involves diagramming conversations I have on the telephone which may cause me distress. I pick a different appropriate color for each person speaking and I draw a line showing the energy in what that person is saying. After a few of these diagrams I usually find that I can bring the level of intensity in the conversation down to a calmer place and diffuse the anxiety. The other person doesn't know of course that I am drawing these conversations. In the end I find them very calming and indicative of where our misunderstandings lie with each conversation. The energy gets smoothed out more and it becomes easier to come to an understanding of differing points of view. You can do this with any conversations that you have that create difficulties for you and people you have issues with.

Free Writing

Free writing is an exercise in writing down free flow thought as it comes to you without planning what you are going to say ahead of time. This illustration was done by a calligrapher - someone who writes in decorative formal styles - in which her writing is very elegant and formal while the ideas expressed are free form and just flow on the page. Try just putting down whatever comes to your mind without thinking about it ahead of time. You might make it your own form like the calligrapher did - just by using a special colored pen or marker for variety.

After each journaling exercise write a few thoughts about how the exercise made you feel. Is there something you would like to add or change?

Closure and Commemorations

Once you start to express your feelings and memories through journaling and abstract art forms, you will find that you become more interested in symbolism and imagery beyond the daily reality of what you can see and touch. One very interesting area of imagery involves archetypes. Archetypes are recurrent symbols or motifs in literature, art or mythology like images of good and evil. You can create your own art pieces with your own versions of archetypes.

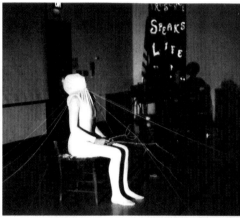

Closure

Sometimes an event in your life inspires you to do something more than just write in your journal. Working with materials to make an art piece is a wonderful opportunity to get in touch with feelings that may be difficult to express in words.

Memorial

A rain painting is a great example of something you can do to take your feelings to another level. I did this rain painting when a friend of mine died. I painted her in lavender lying on the green grass and myself in lavender sitting near her and then I soaked the painting so our colors ran together down in the corner and joined me to her.

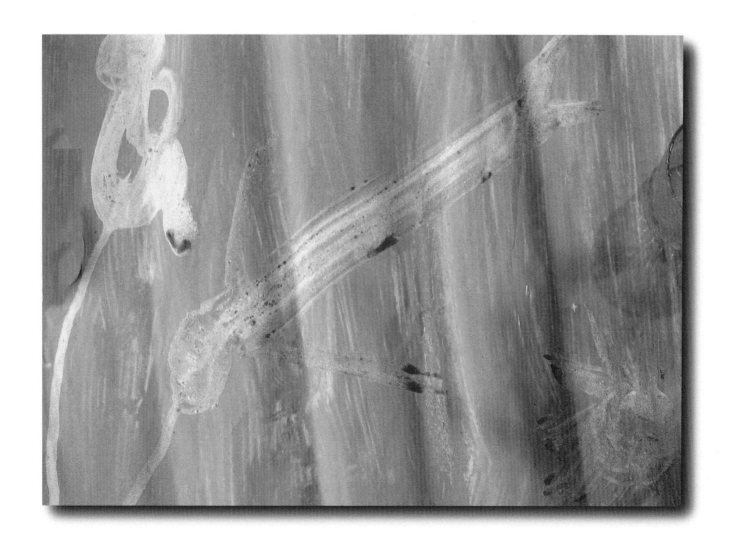

Shelter

Another project involved a sculptural piece made out of metal scraps which I rain painted and wired together like a shelter into which I placed a little garden statue of a fairy representing a young girl who died. I made this piece when a hurricane destroyed my friends house so this piece has 2 meanings for me.

Family Commemoration - *adult supervision on this one*

This family portrait is a collection of hands cast in plaster connected to each other with ribbon in the clouds and stars to bond the family members together. The family dog was so anxious about being left out of the casting that we cast his paw too and included him in the piece. Be very careful to use saran wrap or petroleum jelly, on any surfaces that you cast in plaster so you don't pull anyone's hair out when the plaster dries which takes awhile. If you cast someone's face to make a mask put straws in the nose and mouth so he can breath, and don't forget the saran wrap. Plaster can really pull. This is a nice exercise to do in a group since it involves casting each other and then waiting patiently for the plaster to dry.

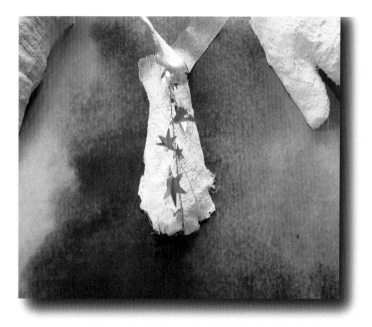
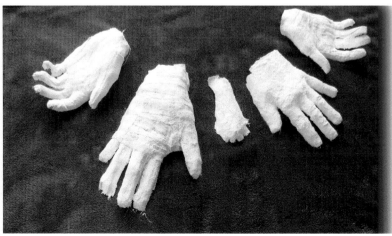

SECTION THREE
Introduction to advanced projects

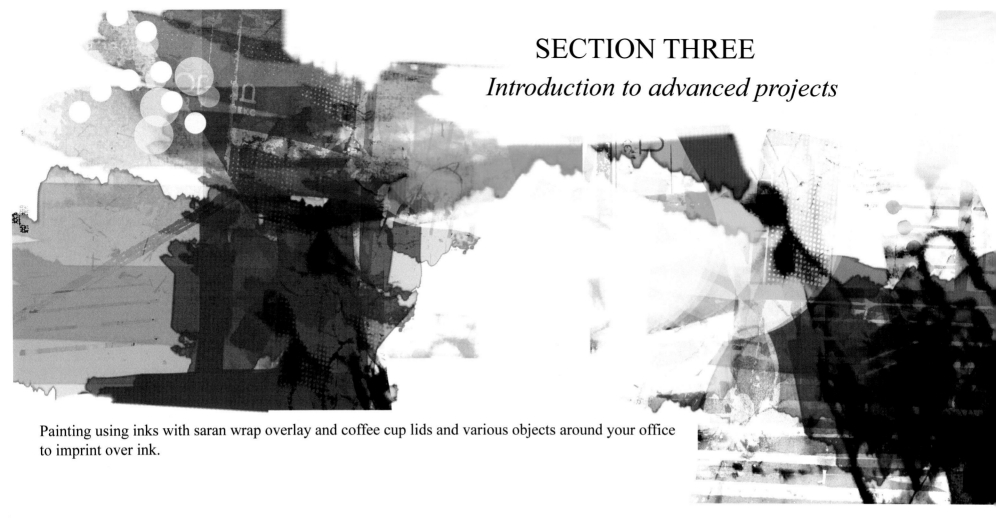

Painting using inks with saran wrap overlay and coffee cup lids and various objects around your office to imprint over ink.

For those of you who wish to explore some of these activities beyond the lesson plans that have been outlined for young people in this workbook, there are wonderful sites on the web and in books where you can explore new techniques. Some of the few listed such as Life Drawing, Collage, Mandalas, Ikebana and Journaling are pursued by artists all over the world in different cultures and using different media than the simple techniques we have introduced here.

We have attempted in the preceding chapters to stimulate interest in creative projects at a very simple level and with easily obtained supplies. As you gain interest in some of these ideas you may want to take the opportunity to branch out and visit galleries, museums, street fairs and studios to observe what other artists do with their materials and ideas. They can be a real inspiration to you as you move forward in your personal journey in the arts and in your life. Just remember that any ideas and inspirations that you have are as valid as anyone else's might be. So get out, go explore and most important of all, have fun.

Indian Rug Collages by Joya Hoyt

Most of my consistent work in visual arts has been in the creation of photo collages on Navajo rug designs. Part of my process involves taking the photographs myself and printing two copies of each so I can do mirror images.

After I decide on a story line or myth for the theme, I begin to assemble the symmetrical patterns following the shapes of the rug patterns but with my own pictures inside the lines. I work with scissors and glue building the photo pieces onto the paper which has been xeroxed with the Navajo rug patterns.

Each RUG has a theme or story or event involved in the pictures I use. Cinco de Mayo, Patagonia, Seahorse, Nantucket Garden, Desert Church, Navajo Silver Jewelry, themes reveal themselves in the collages. All of these pieces have significant family history and personal imagery.

The octagonal pieces are based on antique Sailors Valentines. Each one of these "rugs" tells a personal story about my life.

We created a quilt in the 6'th grade art class that is shown in this book. For that project we used images cut out of magazines which were more readily available than photographs.

Although it is possible to "cut and paste" with computers today, there is something much more satisfying and deeply connected about doing the process with scissors and glue. Building an art piece with tactile involvement adds a personal dimension and sensation to the construction that the remote activity of using a computer lacks.

Close up of rug above

Close up of rug

Close up of rug to the left

Close up of rug to the left

Close up of rug to the left

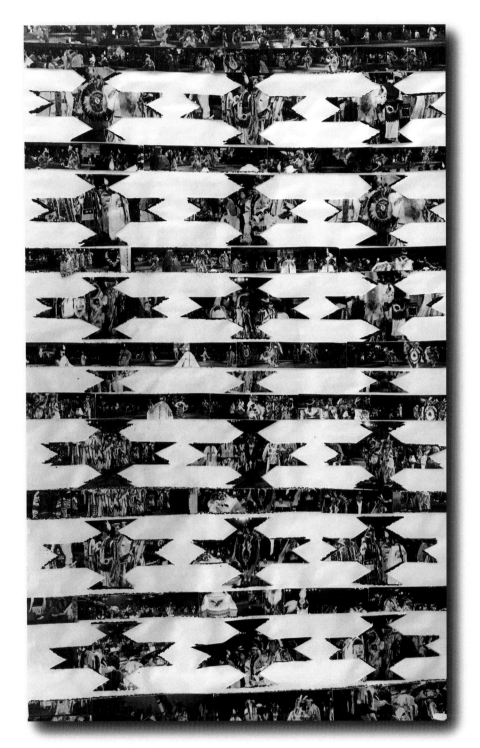

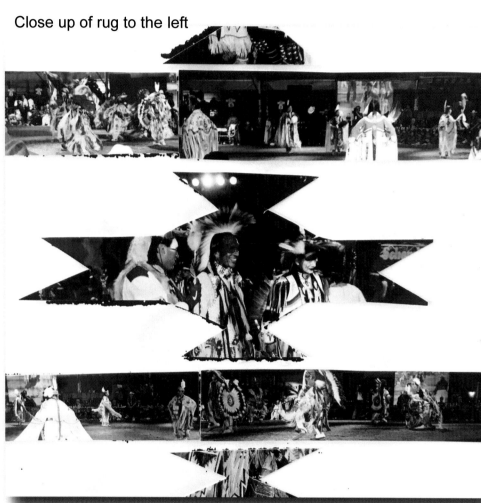

Close up of rug to the left

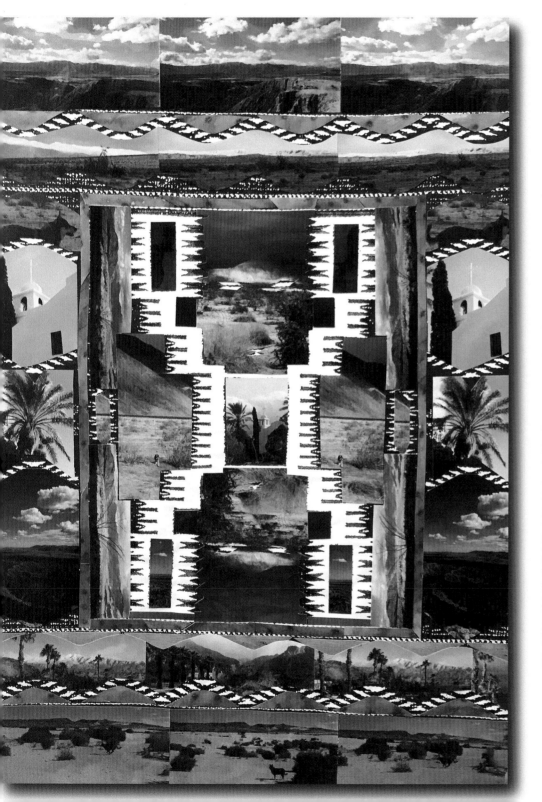

Close up of rug to the left

Close up of octagon rug to the left

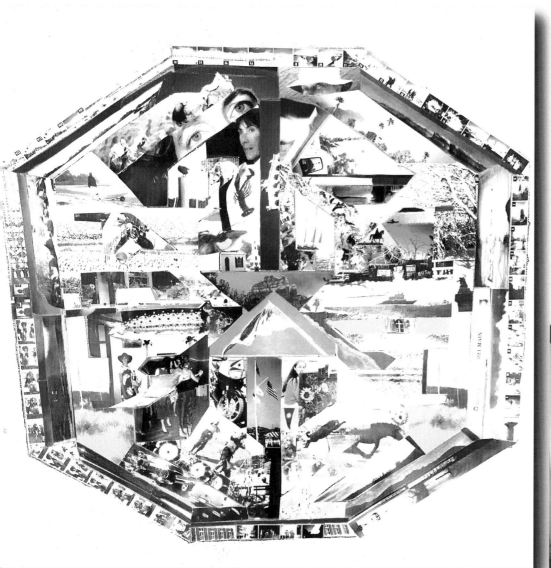

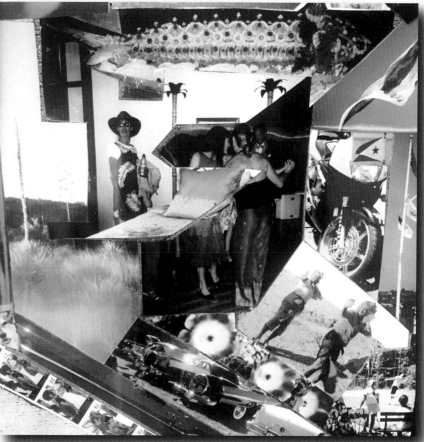

Close up of octagon rug to the left

Alcohol Ink Painting... Journey to your soul

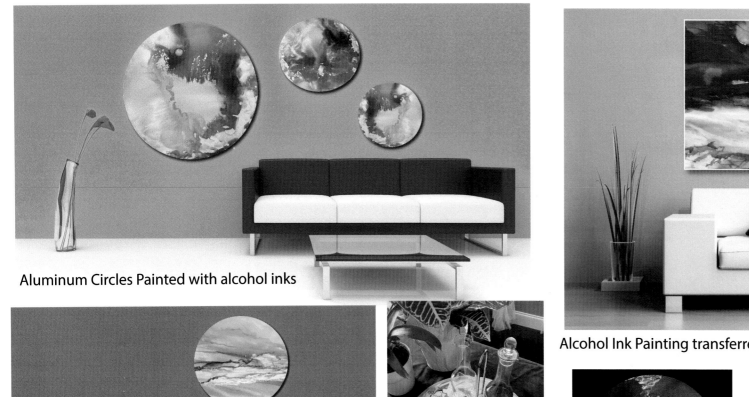

Aluminum Circles Painted with alcohol inks

Alcohol Ink Painting transferred to canvas

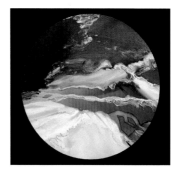

Aluminum Circle made into Lazy Susan

Tiles in frames

Each tile that is painted with alcohol inks takes the painter on a spiritual journey that connects him/her with the depth of their soul. It makes you feel alive. Creating leads to awakening and a feeling of worthiness. You begin to listen to your inner voice; you begin to excavate the wild within; you allow the power of beauty to shine from within. Flowing flowers, a tempest, a deep abstract, a Georgia O'Keefe look-alike, a Monet look-alike, take form with the splash and natural flow of ink.

Frames, trivets, mirrors, trays, napkin holders, bookends can be used to display their creations and use them for decorating their room or to give as handmade gifts for holidays or birthday presents. Some students have even continued on learning to selling their art at farmers' markets or online.

ADVANCED COLLAGE

A collage is a form of visual arts in which visual elements are combined to create a new image that conveys a message or idea.

Collage describes both the technique and the resulting work of art in which pieces of paper, photographs, fabric and other ephemera are arranged and stuck down onto a supporting surface.

How to Make Your Own Paper Collage Art

1) Start with a base for your art. ...
2) Cut or tear strips of paper. ...
3) Layer the paper in stripes on the base. ...
4) Glue into place. ...
5) Add accent pieces, embellish your art to finish it off.
6) Optional – make spacers to place in between paper layers to create some dimension in your art piece.

Collage is an art form that is made up of overlapping pieces of material, such as The process focuses on the act of selecting materials and cutting them into the desired shape, before arranging them and pasting them onto the chosen surface. Unlike other art forms that may rely on proficient technical skills which may require more time to master, such as painting and sculpture, the artistic expertise of collage lies in the choosing, arranging and affixing. As such, the term 'collage' comes from the French term coller, meaning 'to glue' photographs, fabric, colored and textured paper and other types of mixed media.

ADVANCED IKEBANA

IKEBANA – the ancient Japanese art of flower arranging according to strict rules

The basic elements of all ikebana arrangements are color, line, and mass. Ikebana symbolizes the beauty of all components of the natural world working in harmony. You don't just use flowers and greenery but can add twigs, moss, stones, and even fruit.

What does ikebana look like?

Ikebana arrangements are not unlike sculpture. Considerations of color, line, form, and function guide the construction of a work. The resulting forms are varied and unexpected, and can range widely in terms of size and composition, from a piece made from a single flower to one that incorporates several different flowers, branches, and other natural objects.

As an ikebanaist, you trim flowers and branches into unrecognizable shapes, or you may even paint the leaves of an element. Plant limbs may be arranged to sprout into space in various directions, but in the end, the whole work must be balanced and contained. At times, arrangements are mounted in a vase, though this is not always the case.

In ikebana, it is not enough to have beautiful materials if the materials are not artfully employed to create something even more beautiful. You can become a skilled maker by carefully placing one flower and that can be just as powerful as an elaborate arrangement.

Beginners are taught how to sensitize their eyes to the materials, to be able to bring out their inner qualities, and understand how this changes with each arrangement. Beginner arrangements done often make use of two tall branches and a small bundle of flowers. These pieces follow the three-stem system of shin, soe, and hikae—elements that have traditionally represented heaven, man, and Earth, respectively. Now, on a practical level, they refer to the main stems that are employed. All other stems are called jushi, meaning supporting or subordinate stem.

Arrangement made with stones, branches and dead leaves

Christmas minimal style composition. disco ball , tree branch, crumpled paper on natural stone

Credits

Motivational Cards - Angel Cards / Art Therapy Cards / Breta Matson

Diamond Thief - Timothy Elliott

Archetypes Owl - EAT Magazine (excerpt)

Spider - Pamela Granbery

Tigers - Anna Halprin

Beach Mandala - Joan Hall

Calligraphy - Kay Atkins

Canyon - Kerstin Zettmar

Film Cover Art - David Leclerc

Dog Painting - Mario

Sewing Director - Ann Powderly

Cooking Instructor - Pam Levens

Biography

Joya Granbery Hoyt is an educator and performer from Newport, Rhode Island. Joya started out teaching dance to the neighborhood kids when she was ten years old. She performed in an elementary school improvisational theater group at Guilford, Connecticut and in her high hchool dance group. That led to a college major in Theater at Vassar and a Master of Fine Arts in Dance at Connecticut College. She studied dance therapy at the American Dance Festival and performed with the Conetic Dance Theater with Martha Myers and the Connecticut Wesleyan Experimental movement lab and with Anna Halprin.

The ADF Dance Television Program and Community Outreach Dancing in the Streets furthered Joya's expressive arts focus as she acquired a Masters in Special Education at Salve Regina University. She began teaching in the public schools in eastern Connecticut and Rhode Island and continued touring the Rhode Island schools with the RIC Dance Company as well as performing at Clarke Center in New York. During this time Joya founded the Newport Cultural Affairs Commission, which is now The Arts And Cultural Alliance of Newport County.

A grant from the NEA took Joya to California to co-found a dance company called Sisters, performing in Los Angeles and San Francisco. She spent 20 years working in the film industry making features and documentaries around the world and afternoon educational specials for television.

Joya returned to Newport to focus on philanthropy and preservation culminating in a costume show at Rosecliff Mansion where she is a member of a number of local committees and performs with dance companies in New England. In the winter Joya resides in California where she and her sister, Pamela, run a program called Outside The Box The Middle School Science and Art Project which finds volunteer teachers to conduct afternoon programs in an under served desert community. Her graduate degree in expressive arts at Salve Regina University added depth to her work in these arenas. She also enjoys working with the elderly in this field, as well as broadening her own art work from photo collage to other media.